VIKINGS: SHIPPING AND TRADING LESSONS FROM HISTORY

BY: Mustafa Nejem

CONTENTS

INTRODUCTION

Have you ever heard of the Vikings? What images pop into your head? Helmets with horns? Fur-clad warriors? Fierce raiders and pirates? The truth is, the Vikings are more fascinating than you might think.

Contrary to common misconceptions, the Vikings were not a single race or ethnicity. Instead, they were a diverse group of people who lived in the regions of Scandinavia, comprising modern-day Sweden, Norway, and Denmark. Another interesting thing about Viking is that the word Viking has evolved in meaning over time. Originally, it referred to army expeditions or armed raids by Scandinavians. However, today, it is commonly used to describe people from the Viking era, encompassing individuals of all ages, from children to older people. This linguistic shift reflects the broader popular understanding of the Vikings as a historical and cultural group rather than solely as raiders or pirates.

The Viking Age is generally considered to have spanned from the late 8th century to the early 11th century. Many Vikings were farmers or rural settlers who turned to a life at sea due to factors like overpopulation, desire for adventure, or economic opportunities. This transition from farming to seafaring marked a significant historical shift, leading to their exploration, trade, and raiding activities across Europe and beyond. The exact reasons for this shift from farming to seafaring remain unclear. It could have been due to harsh weather conditions, making it difficult to grow enough crops to sustain their families, or perhaps a desire to seek wealth and adventure beyond their homelands. The motivations behind their transition could have been numerous, but the precise cause remains a mystery.

However, what we do know about the Vikings is that they were remarkable warriors and skilled seafarers. They displayed tremendous courage by venturing into unfamiliar waters, and their cleverness was evident in their ability to discover new trade routes, adapt to dynamic markets, and, perhaps most impressively, construct magnificent ships capable of long-distance voyages to trade goods for profit. Their adventurous spirit led them to explore uncharted territories, forging connections with distant lands through trade. The Vikings had a distinctive shipbuilding technique and were highly proficient in sailing and navigating the seas.

Ancient Scandinavia corresponds to modern-day Denmark, Sweden, and Norway, but Sweden did not play a major role in the Viking Age. They often interacted with the Russians and settled in regions such as Finland, the Baltic states (especially Estonia), Russia, and Belarus.

In addition to these countries, the Frisians and Saxons in Germany also interacted with Viking invasions. The main groups the Vikings dealt with included the English, Irish, and Frisians, who often viewed them as robbers and plunderers. However, we will now shift our perspective away from seeing them solely as pirates (while still acknowledging historical facts).

This book will delve into a different aspect of the Vikings. We will examine them through the lens of their shipbuilding and trading endeavours. We'll discover what set them apart and contributed to their success compared to their counterparts in the Middle Ages. We'll explore how they ventured into distant seas and lands through conquest and trade and rose in this industry like a phoenix from the ashes.

In ancient times, raiding and trading required more than just the desire to sail and a spirit of adventure. Numerous factors came into play, especially without advanced technology and awareness. Of these factors, ships played a pivotal role.

The Vikings, renowned for their skills as both warriors and sailors, also excelled in the art of shipbuilding. Their iconic longships were meticulously constructed using a clinker-built technique, where iron rivets held overlapping planks together, resulting in robust yet lightweight vessels. Furthermore, their ship design and architecture were surprisingly advanced for their time. The selection of wood, such as oak and pine, was strategic, and they typically used fresh timber, which allowed for greater flexibility in shipbuilding. If their funeral ships were a testament to their craftsmanship, one can only imagine the magnificence of their warships and trading vessels.

After mastering shipbuilding, the Vikings also excelled in the realm of navigation. They displayed remarkable navigational skills, relying on their ability to read the sun, stars, and familiar landmarks to guide them across the vast oceans.

In addition to their shipbuilding and navigation skills, the Vikings were outstanding traders with a keen sense of market demand. Instead of limiting themselves to a single product type, they traded various goods, including silk, fur, animal skins (especially from whales), and precious metals. They understood the value of luxury items and had a strategy of purchasing abundant products in one region and selling them where they were rare, allowing them to excel in business and generate more profit.

Furthermore, the Vikings were skilled traders adept at establishing trade posts strategically. They strategically positioned these trade posts along rivers and coastlines, demonstrating their deep understanding of trade and security. These locations offered multiple advantages: they provided access to both maritime and inland trade routes, with rivers playing a crucial role in commerce; they served as defensive positions against rival tribes, making surprise attacks more difficult; and they facilitated cultural exchange and the standardisation of trade practices, fostering a stable and conducive trading environment.

Overall, the Vikings' experiences in trade, shipbuilding, and diplomacy enabled them to adapt to the market and establish global connections while preparing for potential threats. Whether establishing the Vinland settlement or forging relationships with the Byzantine Empire, the Vikings prioritised building strong ties with foreign traders, including important figures and royals. This approach protected them from pirate raids and provided them with the security of kings' protection in exchange for taxes and riches. It was a mutually beneficial arrangement. In summary, the Viking era was characterised by trade and remarkable shipbuilding, showcasing their craftsmanship and shrewdness. These lessons hold valuable insights for modern shipbuilders and businesses, which we will explore further in this book.

Chapter 1

BUILDING A UNIQUE BOAT

From colonising Greenland and Iceland to terrorising the monks of Lindisfarne and establishing trade routes for the Norse, Vikings could never have solidified their reputation as warriors and the terrors of the medieval era without their shipbuilding skills. Vikings constructed impressive dragon-headed longships, also known as "drekar," which seamlessly explored and raided various regions from northern England to North Africa. They utilised two types of ships, Knarr and Karve, for both trade and warfare, and these vessels were notably more efficient and lightweight compared to their English and Frankish counterparts of that time.

The information and visuals we gathered from sagas, archaeological findings, and written records have provided insights into what Viking ships might look like. One of the most iconic images associated with the Vikings is the dragon-headed longship, with its red-and-white striped sails that made it incredibly fast. These ships transported fierce warriors to their intended destinations in search of riches. Their unique dragon-like design and robust architecture allowed them to sail through rough waves and strike fear into the hearts of those in various regions during the medieval era. In this chapter, we will delve into the uniqueness of the ships used by the various Scandinavian people known as Vikings, examining their designs, materials, and how the Vikings mastered the shipbuilding techniques of their time.

Description of Viking Longships

The most famous image of a Viking longship is the drekar, also known as a dragon-headed longships. These iconic ships are often depicted in movies and artwork whenever Vikings are portrayed. Maybe it is why a great Viking ship, with a name like "Sea Serpent," would smoothly glide over the crest of ocean waves. While many associate intricately carved dragon heads with Viking longships, not all featured such decorations. In reality, Viking ships came in various shapes and sizes, including large cargo vessels with ample storage and fast longships ideal for raids, providing the Vikings with versatility. Regardless of their specific designs, the remains of these ships show their capability to navigate challenging waters and their sturdiness for long-distance travel. Now, let's dive into the design and archaeological marvels of these ships.

Meticulous Design and Construction Techniques

Here's a fascinating fact. All Viking ships were constructed using the clinker shipbuilding method. In clinker ship construction, they built the outer hull before adding the internal frame. The planks were positioned with one edge overlapping the other and secured in place with rivets. This technique contributed to the durability and distinctive design of Viking ships.

The Vikings created their boats using basic tools, and there's a belief that a Viking boat could be constructed using only an axe. Despite these simple tools, their craftsmanship was quite advanced. They skilfully worked with the wood's natural grain to create strong and flexible boats while keeping them as lightweight as possible. The Vikings had two primary types of ships: Knarr and Karve, both of which could be clinker-built. Knarrs were larger ships designed for longer voyages, while Karves were more suitable for navigating shallow waters. Vikings appreciated the lightweight nature of their boats, which is likely why they opted for clinker construction. Clinker-built boats offered structural strength without needing a heavy frame, allowing forces to be efficiently transmitted between the hull and the means of propulsion, like oars and sails. This made their ships both strong and agile.

The oldest plank-built ship, the Hjortspring war canoe, hails from Denmark and dates back to around 350 BCE. Similarly, the Nydam ships, three vessels found in Denmark, with the largest being approximately 23.5 meters long and 3.5 meters wide, are from around 350 CE. What sets these ships apart is their use of overlapping planks, a departure from the earlier practice of tying planks together. The Vikings introduced another innovation using iron rivets, a key feature of Viking shipbuilding.

It can be that the Viking shipbuilding process began with crafting the keel, which served as the foundation for their boats. They also employed a clever technique known as the "negative keel" to counteract being blown off course while sailing. The ship's shape was largely influenced by the curved posts at the front and back called the stem and stern. The Stem-smith, responsible for shaping the hull, played a crucial role in making planks fit together to create the ship's unique form. It's important to note that while the Vikings didn't possess modern hydrodynamic knowledge, their shipbuilding techniques evolved over generations through trial and error. They didn't rely heavily on detailed plans or measurements but followed practical rules of thumb. All of these practices demonstrated their dedication to crafting finely crafted ships.

Significance of high-quality timber in construction

One of the great examples of the Vikings' success in shipbuilding is their careful choice of timber for their ships. Viking ships are renowned for their high quality due to their clinker design. Unlike other ship designs like carvel, clinker construction requires top-notch timber, typically oak. Vikings excelled in creating ships known for their unique design and exceptional strength, making them well-suited for open sea voyages and long-distance travel. Their commitment to quality set them apart in shipbuilding.

In Viking shipbuilding, they were meticulous about choosing the right timber, especially oak. However, they also utilised pine wood for some parts of their ships. Pine planks had a unique advantage: as they aged, they could bend differently depending on whether the side with bark faced the water or the inside of the boat. This allowed them to shape the boat's curve over time. Pine trees could only yield two planks each, and these planks required shaving down the curved sections to look like proper planks.

Besides this, it's important to note that the entire boat wasn't made exclusively from either pine or oak wood. Both types of wood were chosen for specific parts of the ships. For instance, oak was likely used for the keel, while pine was employed for masts and yards.

Furthermore, Vikings had a unique approach to timber for shipbuilding. Another fascinating aspect of Viking shipbuilding was their use of green timber. Unlike modern practices of drying wood for years before use, the Vikings worked with fresh wood right after cutting it down. Fresh wood is more flexible and easier to work with, which was crucial for crafting the intricate shapes of Viking boats. To preserve the wood's freshness, they stored it in water. This approach allowed them to bend and shape the wood without breaking, making it ideal for crafting their impressive ships. This method likely played a crucial role in constructing the expanded logs in Viking ships. There's evidence of this practice, as parts of a Viking boat were found in a lake, suggesting that it was made using extra suitable wood while waiting for more.

The choice of wood types for Viking ships depended significantly on the planks. Oak trees could be split into numerous planks, while pine trees produced only a few. Vikings often used younger pine trees to create smaller planks. Consequently, just two planks could be made from one pine tree, and these planks required shaving down the curved sections to resemble proper planks.

On the other hand, Vikings could produce multiple planks, even up to 64, from a 200-year-old oak tree. These oak planks were slightly triangular and rough, so they were smoothed down, similar to how pine planks were treated.

In addition to pine and oak, Vikings also utilised another type of timber known as grown timber, primarily for inner frames. This wood was naturally shaped, making it well-suited for shipbuilding. Large, curved branches from this timber were used for the stem and stern posts. Additionally, smaller ships or boats that didn't have holes for oars fashioned rowlocks, known as tholes, from the junction of a branch with the trunk, which provided the strongest part of the wood.

Another fascinating aspect of Viking ship materials is the discovery of readily split wood, specifically chosen for its grain to enhance the strength of the ships. Surprisingly, this type of wood is no longer found today. To create these planks, Vikings would split a log from top to bottom using an axe and then widen the split with wedges until it divided into two halves. This technique highlights the ingenuity of the Vikings in selecting materials that contributed to the robustness of their ships.

In conclusion, the Vikings showcased their intelligence by expertly selecting the right materials for each ship component, demonstrating their mastery of shipbuilding.

Importance of Lightweight Ships:

The importance of lightweight longships is related to their adaptability, allowing them to navigate shallow rivers and open seas easily. A group of warriors in the Viking age was used to navigate seas of the distant land for their journeys. These longships were very famous for their lightweight design and adaptability. These long ships impact many journeys in the shallow rivers.

They use different rigging, ship sailing techniques, and adjustments to make it a valuable craft. This adaptability was a big opportunity for the Vikings because it helped them trade and explore distant lands. So, the lightweight and versatile long ships became a very enduring symbol in the Vikings age, which reflected their innovation.

The Famous Viking Ships the Gokstad Ship

The Gokstad ship, discovered in 1880, is a remarkable example of a Viking vessel from the 9th century. It was found almost fully intact along the shores of the Oslofjord, even though it had been buried for nearly a century. This discovery is a testament to the high-quality craftsmanship and products used by the Vikings in shipbuilding. What's intriguing about this ship is its name, derived from the words Gokstadhaugen/Kongshaug in Old Norse, which means "King's mound."

The Goktad ship was used commercially and was suitable for open sea. Like other standard Viking ships, it was made of oak and served various purposes such as warfare, trade, and transporting people and goods. The Gokstad ship was large, approximately 78.1 feet long and 16.7 feet wide. It featured a unique steering system with a quarter rudder and a part called the "wart" attached to the hull. The ship had 16 tapered planks on each side held together by iron rivets. At the bow, all the planks converged. Additionally, this remarkable ship was designed to accommodate 32 rowers and boasted a large square sail, enabling it to reach speeds of around 12 knots. The ship's mast and rudder were adjustable, allowing it to navigate shallow waters. The wood used for the ship was harvested around 890 AD during the Viking Age. Built during the reign of King Harald Fairhair, it could carry a crew of 40 to 70 men. Overall, its well-thought-out design and ability to handle rough seas demonstrated the sophistication of Viking shipbuilding.

Rokskilde 6

Roskilde 6 is another remarkable Viking ship discovered in Denmark's Roskilde Fjord, alongside eight other ships dating from the Viking era to the Middle Ages. According to dendrochronological studies, it was likely constructed after AD 1025. This ship stands out for its extraordinary length, boasting a keel measuring approximately 32 meters, suggesting it may have belonged to a king or earl. The keel's construction sets it apart, featuring three sections and cleverly shaped joints to stabilise rough waters. Furthermore, the ship's construction was meticulous, utilising up to 8-meter-long oak planks and intricately carved double knees to secure the keelson. This made it robust and showcased the remarkable advancements in Viking shipbuilding skills during that era.

Reviving Traditional Viking Boat-Building Techniques

Modern shipbuilders and mariners continue to be in awe of the marvellous shipbuilding techniques employed by the Vikings. This is precisely why numerous Viking ships are being revived and reconstructed. These stories vividly illustrate how marine archaeologists are dedicated to preserving the Viking spirit by meticulously applying their ancient techniques. Now, let's delve into some remarkable examples of Viking ship revivals.

The Beginning of Reconstruction of Skuldelev 5 Ship

One of the most recent examples of Viking ship revival is the Skuldelev project, which commenced in 2022. After extensive excavations, meticulous preservation efforts, and careful reconstructions, the Skuldelev ships now proudly grace the Viking Ship Hall at the Museum. These ancient vessels had distinct roles during the Viking Age, ranging from large cargo ships used for trade to swift warships and nimble fishing boats. Notably, these ships were warships intentionally sunk in 1070 to block a waterway and safeguard against potential invasions. Each of these ships possesses a unique character, offering valuable insights into Viking shipbuilding techniques. Additionally, as Denmark already boasts the 13th-century cathedral as a UNESCO World Heritage site, the Viking Ship Museum has become a primary attraction thanks to the revival of these magnificent vessels.

Furthermore, among these historic ships, Skuldelev 3 shines as one of the best-preserved examples, built by local Danish craftsmen around 1040. The intriguing Skuldelev 5, dating back to around 1030, possesses a distinct character due to its construction using a combination of fresh and recycled timber, some of which features intricate carvings. Furthermore, a full-scale reconstruction of Skuldelev 5 commenced in 2022, utilising both old and new materials for construction.

This is an excellent example of harnessing the quality materials used in Viking ships, as shipbuilders employ techniques to split 200-year-old oak logs from the Vallø Estate. This process involves a remarkable team of shipbuilders, weavers, sailmakers, archaeologists, and natural scientists. We eagerly anticipate the results of their efforts in the form of a revived Viking ship.

The Revival of Oasberg ship in Viking Ship Museum in Oslo, Norway

It's highly likely that modern shipbuilders are already learning valuable lessons from the magnificent shipbuilding techniques of the Vikings. For example, the Oseberg Viking Heritage project successfully created a replica of the Oseberg ship, originally discovered in Tonsberg in 1904. The ship consists primarily of oak and measures approximately 71 feet (22 meters) in length and 17 feet (5 meters) in width. It was designed for sailing and rowing, featuring 15 oar holes on each side, accommodating up to 30 rowers. With a full crew, the ship could achieve a speed of up to 10 knots (equivalent to 12 miles or 19 kilometres per hour). However, the oars are crafted from pine and bear remnants of painted decorations.

To create the replica, the shipbuilders manually split large oak logs into ship planks, incorporating minor adjustments based on archaeological discoveries from the original ship. This replica demonstrated its sailing prowess during a lengthy voyage to the Roskilde Viking Ship Museum in Denmark in 2015.

Lesson for Modern Traders and Shipbuilders

Undoubtedly, the Vikings were masters of the sea, crafting remarkable ships with unique designs and sturdy materials for long journeys. Similarly, we can draw valuable lessons from the Vikings' adaptability and innovation in constructing new boats. We can ensure top-notch performance and safety by merging modern engineering and technology, prioritising efficient rigging and sails, and conducting thorough testing. Furthermore, reviving traditional Viking boat-building methods through replication and conservation efforts helps preserve their rich heritage. In conclusion, by applying these insights, modern businesspeople and shipbuilders can create new vessels that are durable and efficient, perfectly aligned with today's shipping needs. Let's explore what modern individuals can learn from the Vikings' shipbuilding techniques.

The importance of adaptability and innovation in boat building

The Vikings possessed a deep understanding of the perils of the sea. As ferocious raiders and formidable warriors, they ventured into uncharted waters and lands. To conquer these challenges, they needed ships capable of adapting to the water's capricious nature and innovating to withstand the harshest sea conditions. Enter the clinker ships in a picture, a testament to their ingenuity, featuring an innovative design of overlapping planks that rendered them seaworthy. Overall, their unparalleled adaptability and groundbreaking boat-building techniques allowed them to craft vessels tailored to various purposes and conditions.

The need for Using High-quality Material

Modern traders and shipbuilders should have a keen grasp of the importance of selecting the right materials for specific parts of ships. They should recognise the value of using freshly harvested wood from the method Viking had used instead of waiting for years of drying, which can be time-consuming and energy-draining. Furthermore, they should embrace innovation and experimentation in materials, utilising specific materials for certain ship components to enhance safety and sturdiness, all while striving for the perfect result.

The value of preserving historical techniques for cultural heritage and inspiration

The Vikings were not only innovators but also diligent students of history. Take, for instance, the planking technique, which predated the Viking Age. They took this traditional method and enhanced it with iron rivets, making it stronger and more durable. Modern businesses can draw important lessons from this, emphasising preserving their cultural heritage and a deep understanding of tradition before applying innovation. Moreover, contemporary shipbuilders can find inspiration in the Vikings, recognising that learning from the past can be a wellspring of innovative solutions.

Incorporating Modern Engineering and Technology

The Viking shipbuilding marvel has indeed left a remarkable roadmap for modern shipbuilders, contributing significantly to technological advancements in shipbuilding. One of their most notable innovations was the "dreker," more commonly known as the clinker-built design. As mentioned, this design featured overlapping planks, resulting in a sturdier and more seaworthy structure.

In the same way, modern shipbuilders have harnessed advanced technologies like AutoCAD, enabling precise modelling and simulation to optimise ships for performance and safety. Furthermore, they have embraced cutting-edge materials that provide strength and durability while remaining lightweight—a concept reminiscent of the Viking legacy. With these advancements and the knowledge inherited from the Vikings, modern shipbuilders have become masters of contemporary boat design and construction.

As mentioned earlier, the Vikings have consistently amazed modern shipbuilders with their innovative ideas, from the dragon-headed longships to their use of fresh timber, which diverges from modern shipbuilding techniques. Archaeological evidence in the form of ships dating back nearly a thousand years attests to their unwavering commitment to quality in their industry. Moreover, their ship-burying rituals leave us eager to uncover more mysteries behind their shipbuilding skills. Therefore, there is a pressing need for further research to unveil the lessons that can guide the construction of durable and efficient boats suited to today's shipping needs.

Chapter 2

EXPLORATION AND NAVIGATION

The Vikings were known for their remarkable navigation skills. It was exemplified by a well-known Viking named Leif Ericson. His journey was incredible in showing the remarkable navigation skills of Vikings around the year 1000 CE to North America. Ericson, along with his team, raised to North America and demonstrated proficiency in navigation. He had navigated many unknown territories and waters. This journey of Leif Ericson showed the Viking's expertise in navigation, which relied on natural landmark information, celestial observations, and the mastery of the seas. In this chapter, we will explore Vikings as navigators and their navigational skills, which helped them explore almost half of the world.

Role of Historical Viking Navigation Techniques

Historically, Viking navigation techniques mainly relied on the celestial bodies, including the north stars and the sun. So today, modern navigation instruments like sextans and GPS systems are used for the celestial navigation principles. These techniques provide very accurate positioning of the ships and other vessels in the sea. They are very helpful for both water-based and land-based transportation. Vikings also used natural landscapes in the sea journey.

They were very well known for innovative approaches to navigation. The sun was used as a compass for navigational knowledge. So, by adopting this spirit of innovation, modern businesses can easily improve the different navigation techniques.

Without modern instruments, the Vikings adapted to use natural elements like stars and sun for navigation. The position of the sun helped them in providing valuable information about the direction. Vikings determined the east-west direction based on the sun rising and setting. At night, the stars were used for navigation. They depict the direction of travel by the stars at night along the coastline. The Vikings used natural landmarks for navigation.

These include cliffs, mountain ranges and islands. Moreover, they also paid close attention to the tides and ocean currents. Understanding the flow of currents and tides helped them optimise their voyage and plan their routes. All these navigation techniques allowed the Vikings to explore distant areas of the North Atlantic region.

Viking sailors relied on different constellations, like the North Star, to guide them on their voyage. In the Viking Age, the Vikings used a longship on a moonless night. The crew members started to move against the North Atlantic winds. They mainly focused on the starry sky. There were countless stars, in which one star held their unwavering attention, which was a North Star. They got the idea that as the star was moving with them, they were heading towards the right direction. It was a very intimate connection with the natural world of Vikings. It showed their ability to navigate unpredictable and vast oceans using different stars.

Modern Navigation Tools:

Modern navigation tools are very important for shipping operations. For example, the tools known as Global Positioning Systems (GPS) and electronic chart systems enhance the safety of different maritime operations. These tools provide exact information about the position of the vessel and the vessel's surroundings. It also indicates the potential hazards and risks in the sea. Similarly, the real-time data is also a very efficient navigation tool. It enables the vessels to navigate the storms and avoids delays or damage. These tools are precise in positioning and ensure risk reduction for the ships.

Modern navigation tools also include integrated communication systems to help the vessels stay connected with the other ships' emergencies or show-based authorities. It also allows them to respond quickly in case of any emergency. So, in today's modern world, these navigation tools are very important for the maritime industry. They help contribute to environmental sustainability and also improve the safety and efficiency of the sea operations.

Lessons for Modern Businesses

Modern businesses can learn numerous lessons from the Vikings regarding navigation expertise in many time operations. Vikings were well known for their exceptional navigational skills in trade and exploration of distant land. So modern businesses should also recognise the importance of highly skilled navigators and navigation tools that ensure the efficiency of maritime operations.

Use of Accurate Navigation Tools:

Vikings used accurate navigation tools, which helped them mitigate the risk during sea journeys. Different navigation tools can be adopted in the modern world to prevent accidents and reduce environmental impact. Modern businesses can use the stools for tower operational costs, which makes it a worthwhile investment.

Develop Natural Navigation Techniques:

Vikings were innovative and developed different navigation techniques and tools, including the sun's composition. Modern businesses should also adopt a similar strategy of innovation. Through this, they can use better navigation techniques to stay competitive and operational in this field.

Vikings also combined traditional knowledge with new techniques. Similarly, the modern business should integrate traditional navigation tools with the new technology. These include GPS, digital charts or radars for optimising maritime operations.

The historical success of Vikings in their maritime journeys is a big inspiration for modern businesses; by emphasising navigation expertise, modern business can lead to smooth operations and increased customer satisfaction. It will also increase the reputation of the businesses and modern companies in the maritime industry. They should adopt all the essential tools for efficient and safe sea journeys.

Encouraging Investment for Efficient Shipping

Encouraging investment for efficient shipping is a valuable lesson for modern businesses that can be learned from Vikings. While Vikings were professional seafarers known for their best navigational skills, modern businesses could also learn from them and invest in different technologies. These technologies include satellite communication, GPS systems, or other weather forecasting technology. So modern businesses can invest in enhancing maritime navigation.

Additionally, modern businesses should similarly invest in technology while honouring the legacy of the Vikings. They always relied on innovation for navigating the seas. Modern businesses can also improve safety, fuel efficiency, and route planning to get benefits in sea navigation.

Overall, Vikings used many techniques for navigation in the sea. They used natural landmarks, celestial bodies, seamanship skills, and other innovative ideas for improved navigation. Modern businesses can easily invest in navigation tools by taking lessons from Vikings about navigation techniques and technologies. They should invest in tools for enhancing safety in vessel positioning and weather conditions. Modern navigation technology will also help businesses in on-time deliveries and fuel efficiency. It will give businesses a competitive edge in the market.

Chapter 3

MARKET RESEARCH AND DEMAND-SUPPLY ANALYSIS

The Vikings had been famed for their giant marketplace research and specific demand and supply analysis throughout numerous areas. As discussed previously, they were acknowledged for their seafaring capabilities and warrior spirit, which thrived from the 8th to the 11th century. Their trade routes spanned massive distances, permitting them to collect precious records of numerous markets and commodities. Before embarking on trading voyages, Viking merchants invested incredible efforts in collecting facts about the desires and availability of goods in different lands. Their keen knowledge of marketplace forces has become a key factor in many worthwhile trading ventures. In case you don't know, Market research and demand-supply analysis are vital in expertise purchaser behaviour and making informed enterprise selections. To gain precious insights, you could draw belief from surprising resources.

In this chapter, we will delve into the arena of Vikings to explore their marketing strategies and how their strategies and practices can enlighten modern businesses. What kind of market research have they conducted, and how have they fared? We'll also learn about some well-known Viking traders, such as Olaf, and their roles in market research among the Vikings to have a clearer picture of their times.

Olaf and his trade strategies

Have you ever heard about a Viking dealer, Olaf, who sailed his longship to remote corners of Europe, Asia, and the Middle East in search of precious trading opportunities? Unlike many merchants of his day, Olaf did not depend on assumptions or hearsay. At each new port, he took time to discover the neighbourhood markets, look at what items had been scarce or in excessive demand, and chat with consumers and dealers to learn more about nearby wishes. Olaf returned from his voyages with prised commodities like furs, uncommon spices, and valuable metals—not simply by way of hazards but because his thorough marketplace research allowed him to perceive goods that would fetch maximum earnings.

How did the Vikings collect facts about special areas and their changing trade needs?

The Vikings accrued critical market records from their long-ranging explorations and interactions with numerous cultures. Their popularity as awesome travellers and raiders took them to lands near and far, from Western Europe to the Arabian Peninsula. Such exposure allowed Viking traders to gain first-hand insights into potential new markets and possibilities.

Consider Harald, a merchant who joined several raids along the coasts of modern-day France. During pillaging monasteries, Harald carefully observed the place's weather, herbal resources, agricultural capability, and crafts. He cited that northern grapes and certain pickled foods from his native land were unknown in France. After returning domestically, he organised a successful trading day trip to export those goods to France, earning a good-looking income.

Like Harald, most Viking buyers of the time increased their understanding of marketplace demands and conditions through a vast journey and cultural change. Whether touring bustling ports or joining raiding events, they used every threat to gather knowledge that could find new trade.

Emphasis on Analysing Demand and Supply Dynamics

Viking merchants emphasised analysing the call for and available supply of unique goods in extraordinary regions. By cautiously reading neighbourhood markets, they could become aware of imbalances between supply and demand that signalled potential buying and selling opportunities.

Svend was a prominent dealer who regularly journeyed to markets around the Baltic Sea. As he travelled, Svend noticed that first-rate furs and animal hides were in short supply amongst Baltic communities, in all likelihood due to overhunting. But those items have been abundant in his local Norway. Recognising the demand-deliver imbalance, Svend went back to Norway, loaded his longship with furs, and sailed lower back to the Baltic to trade them for iron, grains, and different Baltic goods scarce in Norway. His insightful evaluation of demand and delivery styles paved the way for a rewarding new direction.

Therefore, astute analysis of supply and demand dynamics became the signature of Viking buyers. By pinpointing goods that have been without difficulty to be had domestically but scarce abroad, they uncovered possibilities for profitable exchanges that benefited both buying and selling partners.

Viking approach to trade goods in excessive demand and restrained supply

Capitalising on their extensive market research, the Vikings advanced a signature strategy of focusing their buying and selling efforts on goods that were scarce in supply but excessive in nearby demand. They recognised that trading such items in poor areas might be extraordinarily worthwhile.

For example, treasured metals like gold and silver were in constant demand throughout Viking change zones, but herbal deposits of these metals were rare in most regions. Viking explorers ventured to remote lands recognised to harbour wealthy metal sources, like Sweden, and returned with valuable metallic ingots and jewellery to alternate. The scarcity and excessive cost of those metals ensured the Vikings ought to charge top-class charges.

Spices have been every other prised commodity in confined supply throughout Europe and Asia throughout the Viking generation. Viking investors found significant income by delivering spices gathered from foreign lands to spice-disadvantaged communities. However, the widespread demand for nearby shortages of products like metals and spices gave Viking investors leverage to obtain huge trading margins.

How did this approach lead to worthwhile buying and selling ventures?

By focusing on domestically scarce items but universally in demand, Viking traders could perform extremely rewarding buying and selling voyages along routes with the Dnieper River to Constantinople and the Volga River to the Caspian and Black Seas.

The Dnieper exchange direction linked the Vikings to the state-of-the-art Byzantine Empire, wherein demand for Viking commodities like furs, timber, honey, and slaves ran high. Alternatively, the Vikings obtained Byzantine gold, silks, spices, and other coveted goods not discovered in their hometown. By capitalising on the unique supply-demand dynamics among those areas, Vikings like Ragnar Lothbrok constructed colossal fortunes through this trading network.

Similarly, the Volga exchange path granted Vikings access to uncommon commodities from the East, especially Arabia and Central Asia. Plentiful Viking commodities like woollens and walrus ivory were exchanged for exceptional wares such as rhinoceros horns, valuable stones, herbs, and spices. This course became so profitable that the Vikings set up outposts along the Volga to control this vital alternate.

Conducting thorough modern market research

The Vikings' meticulous efforts to look at and apprehend target markets are more relevant than ever in the modern, complicated, and rapid-paced worldwide economy. While contemporary agencies have way more direct access to data, simply having statistics is insufficient. Companies should recognise a way to successfully accumulate, examine, and observe market insights, as the Vikings did centuries ago.

Modern multinational agencies can ask Viking buyers for proposals when conducting marketplace studies. Consider IKEA, the Swedish furniture giant with masses of shops internationally. Like the ways-roaming Vikings, IKEA invests heavily in reading the needs, tastes, and cultural nuances of diverse markets before introducing new stores.

For the Middle East marketplace, IKEA conducts giant surveys and consciousness groups to recognise local furniture possibilities and decor patterns. Researchers visit houses and workplaces to look at how spaces are used. This meticulous homework lets IKEA conform product designs and save studies to better cater to local tastes, mirroring the Vikings' localised method.

Meanwhile, groups like Apple constantly survey suppliers, producers, and distributors in key international areas to identify problems or inefficiencies in manufacturing and distribution chains, much as Vikings monitored commodity flows throughout exchange zones. Such vigilance helps Apple pinpoint untapped opportunities or deliver bottlenecks before competitors.

Lessons for Modern Businesses

Modern corporations can draw key lessons from the Vikings' pioneering use of specified marketplace research:

- Assess target areas very well before entry. The Vikings accrued on-the-floor intelligence at superb length earlier than embarking on new change routes. Firms want in-depth knowledge of nearby political climates, purchaser needs, supply chain troubles, and different market-unique dynamics before investing in a nearby enlargement.

- Focus on high-call for scarce-deliver goods. The Vikings identified items that were prised and tough to achieve regionally for their incredibly worthwhile buying and selling. Similarly, firms must focus their portfolios on services that tap into unsatisfied regional demand. Unique products or business fashions can achieve success in underserved markets.

- Adapt items and routes primarily based on changing conditions. The Vikings continuously changed routes and traded goods to capitalise on moving supply-call dynamics. Regular marketplace re-assessment lets modern companies pivot services nimbly when situations evolve.

- Another key lesson from the Vikings is examining purchaser demand tiers and enterprise supply availability in goal markets. The Vikings' near-tracking of nearby demand-deliver patterns allowed them to optimise trading operations and maximise income. Some relevant techniques include:

- Regularly survey customers and groups to gauge calls for shifts for key product categories and forecast wishes.

- Closely related are competitive supply and availability situations for existing services and potential new launches.

- Identify supply-demand mismatches that present gift opportunities to introduce well-matched products.

- Be prepared to ramp up manufacturing and logistics to capitalise on spikes in customer pull and the scarcity of opportunity elements.

- Shift offerings and increase strategies to areas where patron demand surges but delivery is confined.

- Applying Viking training to modern commercial enterprise techniques for successful buying and selling

Here are some real-life examples of profitable market research:

Louis Vuitton: The luxury brand conducts exhaustive qualitative nearby studies to create vicinity-precise product lines and marketing campaigns that resonate with target consumer aspirations and tastes.

Samsung: The tech corporation considerably surveys key B2B markets to pick out unmet needs of businesses and perfect product positioning. Additionally, it continuously runs cease-consumer recognition businesses.

Coca-Cola - Via a huge community of neighbourhood outposts, the beverage titan gathers hyper-neighbourhood client insights across cultural nuances, consumption behaviour, and rising traits in diverse markets.

Inditex, the speedy fashion retailer behind the Zara video display, units runway style and street style worldwide to discover unexpectedly changing client tastes. It then acts fast to translate the trends into affordable business collections.

The importance of adapting to changing call-for-supply patterns in special regions

Another lesson present-day corporations can take from the Vikings is the agility to evolve to shifting marketplace conditions. The Vikings excelled at editing exchange routes and centring goods based on changing wishes in distinct areas.

For instance, while the call for walrus tusk ivory declined in Persian lands, Viking merchants shifted to trading extra furs, slaves, and amber alongside the Volga. When the Byzantine Empire's silk manufacturing faltered, lowering European deliveries, Vikings answered by uploading better volumes of Chinese silks to promote to eager European clients hungry for this scarce, luxurious fabric.

Today's companies want this identical sensitivity to local market evolutions and nimbleness to pivot their services and operations for this reason. MNCs like Unilever often modify product availability, capabilities, and pricing throughout their global markets to align with market fluctuations and customer traits. This potential to respond quickly to delivery for shifts learned from the Viking playbook helps contemporary firms acquire market proportion and earnings goals.

In conclusion, it is critical to inspire modern-day agencies to adopt Viking awareness to gain achievement. By emulating Viking traders' diligence and versatility in getting to know various markets and catering to localised demand, today's companies can become extra-astute worldwide operators. Those who fastidiously discover goal nations, pick out unmet needs, pivot nimbly to marketplace shifts, and are aware of in-call offerings can extract excellent profits from global exchange. In our dynamic international financial system, organisations need the pioneering market acumen of historical Viking investors to reap durable success across borders. Modern businesses should closely monitor local demand and supply dynamics, alter techniques, and focus on satisfying unmet needs for profitable growth.

Chapter 4

ESTABLISHING TRADE POSTS

The Viking trade posts were strategically located along the river banks and coastlines. It showcases the Viking's understanding of trade dynamics and security. The establishment of Hafnir is a remarkable example of Vikings' strategic thinking in the past. It was positioned near the river mouth. The location was a masterpiece of Viking trade strategy. By choosing to settle at the end of a river or the sea, they got many advantages.

Firstly, this strategic location provided them access to maritime and inland trade routes. This is because the rivers were very important commerce highways, allowing goods to be transported deep into the continent. So, the proximity to the sea helped connect with distant land and enabled the Vikings to engage in long-distance trade.

Secondly, this river mouth location also helps the natural disences. Vikings were aware of the potential threats of the rival tribes. So, placing Hafnir near the river entrance allowed ships to move in and out easily. It can make the surprise attacks challenging for the rivals. Also, the surrounding terrain can be used as an advantage in the defensive strategy.

The Viking trade posts were very important in controlling commerce during the Viking Age. People from diverse regions interacted by sharing technology, ideas, and languages. This mixture of culture and ideas enriched the Viking world and contributed to a better understanding of the diverse cultures. Viking trade posts also fortified the traders with a great sense of security and safety. These trade posts imposed many rules and regulations to standardise the practice and disputes over the trade posts. It helped in creating a more stable trading environment in the Viking Age.

Role of Trade Post in Facilitating Trade

The Viking trade posts were crucial in increasing trade efficiency during the Viking Age. It helped foster cultural exchange and facilitated commerce. One of the most prominent anecdotes is that Nordhavan justifies the network of Viking trade posts strategically positioned along various major trade routes. For example, the Viking trade post Nordhavan was at the centre of different trade routes in northern Europe. Its location helped the merchants and traders from different regions. Nordhavan was strategically placed and was easily reachable by sea and land routes.

This easy accessibility helped the traders and merchants travel through different transportation modes. This place also became a centre for cultural exchange. This is because the people from different regions share their technologies, language and ideas at this place for a better understanding of the world. The Viking trade post also provided a sense of safety for the traders. Moreover, the security encourages the merchants and traders to enhance the overall efficiency of trade.

Considerations for Establishing Modern Trade Posts or Warehouses

In this modern world, the modern trade warehouses and trade posts serve as very important logistic hubs. They are spiritual places to facilitate the distribution and movement of goods, which is mainly important in the age of e-commerce, where timely delivery to the timers is very important. Warehouses play a very important role in inventory management for different businesses. They help companies manage inventory levels, pile goods and respond to the fluctuating demand for the goods with their exponential online business growth. The modern trade post has storing, shipping and packing features for the products online, which provides timely and accurate delivery to the customers. The modern trade post provides an increased focus on sustainability. They help to incorporate eco-friendly measures such as ways to reduce energy, efficient design and renewable energy sources. This helps in minimising the environmental footprint.

Strategies for Selecting a Strategic Location:

Selecting the strategic locations for the trading post or the warehouses is a very important decision for businesses. The main strategy includes access to transportation infrastructure. Vikings always ensured efficient outbound and inbound goods transportation. Many businesses opt to choose locations that are near the transportation hubs. By thoroughly analysing the market demand and customer behaviour, you can easily produce higher efficiency. These data-driven insights can help you make the right location decisions and identify the growth potential.

Vikings used to access the availability of experienced and professional labour in choosing skilled labour. An experienced workforce is very important for the efficiency of warehouse operations. Vikings ensured labour availability at their trade posts.

They are also used to evaluate the real estate and land costs in different regions by balancing the maintenance cost and requirements of all facilities with potential business benefits. One of the more beneficial strategies was to research the local regulations and tax laws. All these aspects significantly impact the cost of business in any specific location.

Vikings also evaluated potential risks like political instability, natural disasters, or supply chain vulnerability. So, diversifying the locations helps them mitigate the risk, which was a common strategy they adopted. Modern businesses can also adopt this strategy to reduce the risk and increase benefits.

If we consider Amazon, a global E-Commerce marketplace, it is strategically positioned to enhance trade efficiency and delivery timing. Amazon has placed its full payment centres in major populated areas. They have set up multiple centres in Northern cities like Los Angeles and New York, so one of the primary concentrations is the proximity to the target customer. If you are close to the major population, it reduces shipping distance, leading to lower shipping costs and quick delivery centres near cities, exemplifying this strategy.

By planning for the future growth of the company and its scalability, it is essential to choose a location that accommodates the population areas. Talking about Amazon, the fulfilling centres of Amazon are placed near the specific and right areas that provide the effectiveness of the business.

If we consider Amazon, a global E-Commerce marketplace, it is strategically positioned to enhance trade efficiency and delivery timing. Amazon has placed its full payment centres in major populated areas. They have set up multiple centres in northern cities like Los Angeles and New York, so one of the primary concentrations is the proximity to the target customer. Suppose you are close to the major population. In that case, it reduces shipping distance, leading to lower shipping costs and quick delivery centres in nearby cities, exemplifying this strategy.

By planning for the future growth of the company and its scalability, it is essential to choose a location that accommodates the population areas. Talking about Amazon, the fulfilling centres of Amazon are placed near the specific and right areas that provide the effectiveness of the business.

Lesson for Modern Businesses

The lesson for modern businesses has drawn inspiration from Vikings, underscoring the importance of strategic trade post-placement. It helps in optimising distribution and gaining a competitive edge in the market. Vikings strategically positioned the trading post and many key locations of waterways. It helped access diverse markets. So modern businesses can easily consider the proximity to the target markets and customers, which will help them boost the economy.

Management of Trade Routes:

Vikings also efficiently managed trade routes and networks, which helped them in successful commerce. Modern businesses should adopt sophisticated practices to ensure smooth operations and timely customer delivery. For achieving seamless operations, strategically placed trade posts are very important.

Responsive to Threats:

As Vikings quickly respond to every trade opportunity and threat, modern businesses can also gain a competitive edge by being responsive and quick. The strategic location of the trading post helps the businesses meet customer demands and leads the competitors in timely delivery.

Seasonal Conditions:

Vikings adapted their trade strategies to the seasonal conditions as well as to the geographical conditions they encountered. Similarly, the modern business should tailor the trade post placement based on environmental and local factors.

Long-Term Vision:

Vikings were involved in the trade for long-term gains. So, modern businesses should adopt a similar long-term vision while selecting trade post locations. It ensures that the strategic decisions contribute to sustainable and long-term growth.

Data Driven Decisions:

Vikings always relied on knowledge and experience. Modern businesses can also leverage data analytics for complete, informed decisions. Data-driven strategies are very helpful in optimising inventory management and resource allocation.

Famous Viking Strategic Trade Posts

There are many historical examples and archaeological evidence that provide insight into the strategic trade post locations by the Vikings in the Viking Age.

Hedeby was located at the southern end of the Peninsula. It was one of the largest Viking trade posts in the past. Considering its position near the major overland and the maritime trade route, it connected the Baltic Sea region and the sea. It was strategically placed for controlling trade between Eastern Europe, Western Europe, and Scandinavia. This is one of the vital examples of showing the strategic trade post location of the Vikings in the early age, which helped them increase trade and efficiency in the area.

Vikings also established Dublin as a trading post. It was established in the 9th century, and the location was along the east coast of Ireland. This strategic location allowed easy access to the trade routes to Britain, the HheSea, and Continental Europe. It was considered a major trading hub at that time, facilitating the exchange of goods. Goods include various commodities, including silver, slaves, copper, metal, and other valuable commodities.

Kaupang, Norway, is also a real-life example of a strategic trade post location by the Vikings. It was a thriving Viking trade post strategically placed along the Vestfold region in the 8th and 9th centuries. It connected Scandinavia with the western Europe. This place was known for trading textiles, luxury items, and ceramics, which helped trade in the Viking Age. It is present-day Larvik in Norway, which is also helping modern businesses to succeed by following the Viking's principles and strategies.

Vikings also established Jorvik as a major Trade Centre in the 9th century. It was positioned along the river house, which allowed Viking traders to go into England and connect the trade routes held in connecting trade route between the North Sea and the Irish Sea. The Vikings strategically placed this trading post, and was known for its craftsmanship, including Textiles and jewellery.

Bergen was another important Viking trade post in the past that was located on the western Coast of Norway. It provided access to trade routes, including the North Atlantic, British Isles, Iceland, and the Faroe Islands. This version's strategic location helped make it a hub for exchanging fish, dried cod, and other items. It highlighted the strategic placement of Viking trade posts, which helped in commerce and facilitation of trade in the Viking Age.

Birka is also another example of a strategic trade post. It was situated on the island of Bjorko. It was a prominent Viking trading centre that served as an important link between the Western and Eastern trade networks in the Baltic region. The location of this Birka trade post helped the traders control the trade of goods like iron, fur, amber, etc. Its strategic location helped the traders exchange goods and highlighted the strategic placement of Viking trade.

As concluded, Viking's trade post played a pivotal role in commerce and trade facilitation. They were strategically located along major trade routes, emphasising the importance of location in facilitating trade. They adapted to customs and cultures that highlighted the significance of cultural sensitivity in trade. Viking trade posts were the centre of exchange for a wide range of goods. They established expensive trade networks which underscore the importance of building relationships. So, incorporating the strategies of sustainability, data-driven mechanisms, cost efficiency, and proximity to markets is essential for modern businesses to stay competitive and responsive.

Chapter **5**

FAIR TRADE AND TRUST

The Vikings, frequently represented as powerful warriors who voyaged across battles and navigated dangerous waters across foreign lands, possessed a less celebrated but equally great legacy: their reputation for honest trade practices and the cultivation of consideration in global commerce. The Vikings established themselves as skilled dealers and purchasers who valued honesty, integrity, and fairness in their dealings in addition to their military power.

In this chapter, we delve into the exciting world of Viking truthful trade and agree-with-building practices, uncovering the lessons that present-day agencies can draw from the annals of records to navigate the complexities of trade in today's international landscape. From transparent agreements to truthful pricing, the Vikings' commerce technique is a long-lasting testament to the virtues of moral change and trustworthiness.

Viking Reputation for Fair Trade Practices

In the bustling international exchange and trade, the Vikings were now recognised for their seafaring prowess and unwavering dedication to honest dealings and fair pricing. What made the Vikings stand out in international trading changed their reputation for integrity. In a technology considered a rare commodity in exchange, Viking traders had been ideal for their honesty and equity. They have become trusted companions globally, where deceit is frequently the norm. They established trade towns and used weights to ensure everyone paid enough silver for the goods they were buying.

The Vikings' fair exchange practices were not confined to their fatherland; their voyages took them far and extensive, from the coasts of Scandinavia to remote lands. They committed to moral exchange wherever they went, setting a standard others should best aspire to.

The importance of agreements and contracts

The heart of the Vikings' fulfilment in fair trades lay in their meticulous agreements and contracts. These files served as binding agreements that laid out the change terms, ensuring that both parties understood and agreed upon the phrases. This dedication to written agreements fostered consideration among buying and selling partners, as it provided a clear framework for their interactions. They established trade towns and used weights to ensure everyone paid enough silver for the goods they were buying.

The Vikings were no longer content with mere verbal agreements or free understandings. They recognised the significance of putting the entirety in writing, leaving no room for ambiguity. This practice was a testament to their dedication to transparency and trustworthiness. These agreements are now not one-sided; they result from negotiation and compromise. Both parties had a say in the phrases, ensuring the trade became truthful and mutually beneficial. This collaborative agreement approach helped construct belief by validating a commitment to fairness.

The Vikings' emphasis on distinct agreements wasn't just a formality and cornerstone of their successful trading partnerships. These agreements created an experience of protection and predictability in an alternate, which, in turn, recommended lengthy-term collaborations. The Vikings understood that constructing beliefs through agreements was the key to sustainable and mutually beneficial relationships. An instance of the energy of these agreements may be seen in the dating between a set of Viking traders and a foreign service provider. Initially, a language barrier and cultural differences would have hindered trade. However, via the cautious crafting of agreements that spelt out the phrases clearly, both events located an unusual floor and set up a partnership that lasted for years.

Prioritising Fair Trade Practices and Transparent Agreements

Fair trade isn't always a noble idea; it is a realistic technique for undertaking enterprise. Prioritising equity means treating your buying and selling partners with respect and integrity, which can cause better outcomes for all parties involved. This segment emphasises the significance of fairness in change.

The Vikings didn't prioritise equity just because it was the right thing to do. They identified that, in the long run, fair alternate practices brought about more solid and worthwhile partnerships. Popularity for honesty and integrity attracted extra buying and selling possibilities and opened doorways to new markets. Transparency is the muse of trust in modern-day enterprise simply because it changed for the Vikings. This section explores techniques for creating obvious agreements without room for ambiguity. A clean and transparent agreement is step one closer to building trust with your buying and selling partners.

In the modern virtual age, numerous tools and technologies are available to ensure transparency in agreements. Electronic signatures, blockchain technology, and online contract control structures can all contribute to growing agreements that aren't only obvious but also without difficulty and enforceable.

Lesson for Modern Businesses

Emphasising the cost of honest trade practices in building trust with buying and selling partners.

Building belief is more critical than ever in today's competitive global commercial enterprise. Emphasising honest change practices is not only a moral obligation but a strategic benefit. Businesses prioritising equity in their dealings are likelier to establish a lasting relationship with their trading companions.

The classes from the Viking era are a testimony to the long-lasting energy of accepting as true as alternate. By prioritising honest alternate practices, corporations can differentiate themselves in a crowded marketplace and appeal to partners who value integrity.

Encouraging the use of transparent agreements to set up long-term collaborations.

The instructions of the Vikings enlarge the present-day commercial enterprise panorama. Encouraging transparent agreements is essential for corporations seeking to construct, accept, and foster long-term collaborations. Clear agreements reduce misunderstandings and disputes, making it easier for both parties to work together efficiently.

In a state-of-the-art, globalised, and interconnected world, trading partners can come from various backgrounds and cultures. Transparent agreements offer a commonplace language that transcends those differences, making it less complicated to build consideration and cooperation.

The function of trust-constructing measures in fostering beneficial relationships

Trust-building measures, consisting of constantly turning in on promises and maintaining open traces of communique, are essential for fostering useful relationships. Modern corporations can learn from the Vikings' dedication to belief-constructing practices and apply them to their trade partnerships.

A dedication to belief-constructing measures goes beyond the preliminary settlement. It involves ongoing efforts to nurture and give a boost to the partnership. Regular communication, comment mechanisms, and a willingness to conform to changing instances all construct trust.

Modern organisations that have effectively carried out truthful exchange standards in their commercial enterprise operations

In current commerce, several cutting-edge agencies have wholeheartedly embraced the principles of truthful trade, aligning their business organisation operations with ethical, sustainable, and equitable standards. These forward-thinking companies stand as living examples of ways the standards of equity and agree with, akin to the ones practised by the Vikings in their trading endeavours, continue to thrive in the ultra-modern global business panorama.

- One such instance is a socially responsible corporation that has made fair trade a middle part of its business version. By sourcing products from ethical suppliers and paying honest wages to employees, they have won a loyal patron base and hooked up with consider-primarily based partnerships with their providers.
- Fair Trade USA is a certified organisation that adheres to rigorous social, environmental, and monetary standards. By ensuring fair wages, safe operating situations, and sustainable practices, they have become a beacon for transparency and are accepted as true in supply chains.
- Renowned for its dedication to environmental and social responsibility, outdoor garb agency Patagonia exemplifies the fusion of earnings and purpose. Their transparent delivery chain practices and support for diverse moral projects underscore their willpower to uphold truthful change standards.
- A top-notch enterprise, Eileen Fisher has made sustainability and ethical practices critical to its enterprise model. By championing honest wages and environmentally responsible production, they encompass the ideas of trustworthy change.

These cutting-edge businesses aren't only catalysts for change but examples of how trustworthy alternate requirements can be efficiently woven into present-day trade. Through these real-life examples, we can see how honest exchange practices have converted normal business partnerships into belief-based total collaborations. They now encompass the lessons from Viking practices and contribute to a more equitable and sustainable international marketplace.

In concluding our exploration of the Vikings' chapter on truthful trade and consideration, we discover ourselves enriched with insights beyond time and vicinity. The Vikings, renowned as warriors, would possibly have been equally skilled as traders of integrity. Though often overshadowed by memories of conquest, their legacy shines as a beacon of undying standards that resonate inside modern-day business internationally.

The Vikings' commitment to honesty, transparency, and agree-with-constructing in exchange is a testament to these virtues' long-lasting value. Their meticulous documentation of agreements, honest pricing practices, and unwavering adherence to guarantees stand as a blueprint for moral commerce, which has the energy to set businesses apart in a competitive and interconnected international market.

As we look to our destiny, let us remember the training from Viking technology. Prioritising fairness, creating obvious agreements, and fostering acceptance as true can guide companies towards, at the same time, useful partnerships that stand the test of time. The Vikings' legacy reminds us that, in an ever-evolving world, the undying virtues of integrity and acceptance remain beneficial for achievement, ensuring a brighter and more prosperous future for all who include them.

Chapter 6

DIVERSIFICATION OF TRADING GOODS

We admire the Vikings for their impressive shipbuilding skills and navigation prowess. We also appreciate their clever market strategies and trading techniques, leading to significant profits. But what exactly were the Vikings trading for?

Imagine over a thousand years ago when Vikings invested heavily in shipbuilding and took great risks to explore unknown lands. What kind of business were they aiming for, and what goods did they plan to trade? Scandinavia was vast, offering many resources, but no country could be self-sufficient. Vikings sought riches, textiles, spices to enhance their bland food, precious metals, and, most significantly, slaves, as the market demanded them then. This diversity of goods enabled them to thrive in trade without risking an imbalance in supply and demand. In this chapter, we'll delve into the types of goods the Vikings traded and how diversification helped boost their economy.

Range of Goods Traded by the Vikings

The Vikings traded a wide array of goods, from timber to silver, weaponry to silk, and voyaging as far as Baghdad. Their aim wasn't solely profit but also acquiring precious items that were scarce in their homeland. As discussed in Chapter 3, their smart market strategy involved buying abundant items from one place and selling them where those goods were lacking, leading to increased profits. During their travels, the Vikings traded, exchanging goods like honey, tin, wheat, wool, timber, iron, fur, leather, fish, and walrus ivory. Along their journeys, they also took part in the slave trade, buying and selling captives. Let's now explore what the Vikings traded and what they traded for.

Global Export and Import

The business-savvy Vikings understood the importance of global trade for making profits. As a result, they embarked on journeys to distant seas to acquire valuable products. They sailed across the Baltic to reach Norway and Sweden, journeyed along Eastern European rivers, and ventured deep into Russia. They traded at various posts and later returned to Scandinavian trading towns with goods acquired through these exchanges during their travels. However, it's important to note that they were not only importers; Scandinavia had abundant resources to offer. As local Scandinavians, the Vikings promoted their exports and profited from their region's resources.

Therefore, the Vikings engaged in imports and exports, bringing various goods from other regions to Scandinavia and exporting items from Scandinavia to other parts of the world. This trade network played a crucial role in their economy and influence across different regions.

Glass and Amber

The Vikings had a strong appreciation for glass as well. They skilfully crafted glass into beautiful beads, using them as decorations for various items. Important production centres for these glass beads included Åhus in southern Sweden and Ribe in southern Denmark. These beads held special significance for the Vikings, as they used them to adorn their possessions and make them more appealing. Interestingly, while some towns like Kaupang in Norway imported glass items, others like Ribe in Denmark and Birka in Sweden specialised in producing or trading glass beads. Much of the raw materials for glass and glass items came from Hedeby. In addition to beads, glass was also used to create containers. Viking artefacts have revealed drinking cups with a cone-like shape and small jars with rounded bottoms, showcasing the versatility of glass in their culture.

In addition to glass, amber played a significant role in Viking culture as both beads and decorative stones. However, amber ornaments were primarily created for export from Scandinavia because amber was abundant along the North Sea and Baltic coasts. Scandinavia had a thriving jewellery-making industry that utilised amber, glass, and jet. A notable example is the largest collection of amber artefacts from the Viking Age in Northern Europe, Wolin, Poland. This collection comprises over 20,000 pieces of amber, which appear to have been meticulously shaped and crafted, primarily for making beads and pendants.

Precious Metals and Jewelry

Precious metals and jewellery undoubtedly held a luxurious appeal and were lucrative trade items. While jewellery was a part of Viking culture, it's unfortunate that many common Vikings couldn't afford jewellery made of precious stones or metals like silver and gold. The artefacts of Viking jewellery reveal that most of it was crafted using bronze and even wood.

However, let's focus on the business aspect. Gold and silver were not naturally produced in Scandinavia, so Vikings primarily acquired them through various means, such as raids on different lands or churches and trade. While there isn't extensive evidence about gold, there is clear documentation of the Vikings looting and trading silver extensively. Much of their silver came from the Middle East. For instance, they discovered the renowned Spillings hoard on the Swedish Island of Gotland, which remains the largest one found to date. This treasure trove weighed over 67 kilograms (148 pounds) and contained more than 14,000 silver coins from the Islamic world.

Silver was indeed a profitable commodity for the Vikings. When they needed smaller amounts, silver items were often divided into fragments known as 'hack silver.' This practice allowed them to use silver flexibly in trade and transactions. Moreover, in terms of trade, the silver could be shaped into bars and blocks or exchanged as jewellery. The upside of this trade is that they have introduced the coinage system instead of relying on a barter system.

Textiles

Vikings primarily dealt with two types of textiles: silk, which they imported, and fur, which they exported. The Vikings had a taste for luxury and used silk to display their wealth and opulence. Silk was a highly prised item, mostly imported from China and Byzantium via the Silk Road, named after this valuable fabric.

On the flip side, when it comes to fur, the Vikings sourced it from various animals such as foxes, otters, beavers, bears, pine martens, and, notably, sheep, which were abundant in Scandinavia due to its cold climate. While fur was readily available locally in Scandinavia, it became a highly profitable export item for the Vikings. What made it even more appealing was the Vikings' exceptional weaving and spinning skills, which added to the luxury and value of Scandinavian fur products.

Besides fur, the soft feathers from large sea ducks like eiders represented another luxurious and highly sought-after item during that era. These down feathers provided exceptional comfort and warmth when used for bedding at night. Fortunately, eiders were found in abundance along the coast of Scandinavia. While they primarily used this textile locally, it was rarely exported from the region due to its scarcity elsewhere.

Spices

Spices held a significant place in trade throughout history, regardless of the era or civilisation. The Vikings were no exception, engaging in spice trade substantially. While they could produce spices like thyme, caraway, and mustard in their own country, they lacked an abundant supply of key spices that added flavour to their food. Notably, cinnamon, a product of the hottest climates, was among these sought-after spices. Most of the spices they traded came from East Asia or the Islamic world and played a crucial role in enhancing the flavour of their cuisine.

In addition to enhancing flavour in cooking, spices also played a role in preservation, making them valuable commodities. This is why salt was a profitable trading good for the Vikings, as it served as a crucial preservative in many cultures.

Mercenaries and weaponry

It's essential to remember that the Vikings were a formidable force themselves. They were renowned for their ruthless reputation, often working as mercenaries and raiders for payment. Additionally, there are numerous instances of Vikings serving emperors or kings, either for the wealth they could acquire or for the permission to engage in safe trade. Their military prowess and willingness to serve made them sought-after mercenaries and collaborators in various regions.

Besides human resources, weapons were a significant commodity in the Viking trade. Among various types of weaponry, swords held a special place and were imported in large quantities for Viking raids, as the Vikings had a high demand for such products. They particularly cherished and highly valued Frankish swords, fostering a fruitful business relationship with Frankish traders. However, this trade took a hit when Frankish Emperor Charles the Bald banned the export of a specific type of Frankish sword known as the Ulfberht sword during the latter part of the 9th century. The reason behind this was that Viking raids on Frankish society involved using the same swords crafted by skilled Frankish blacksmiths. Nonetheless, this trade persisted in secret, with archaeologists discovering over 144 swords in Viking-era graves, primarily in locations such as Norway and other parts of northern Europe.

Slaves

Leaving aside all the lists of Viking trading goods, it's unfortunate that slavery was the most profitable product of the Viking era. During their pillaging and raiding expeditions, the Vikings would capture many non-disabled individuals, both men and women, to be taken as slaves. Initially, they would demand ransom from the relatives of those they had captured. If no ransom were paid, these individuals would be exported and traded as slaves, marking a dark and profitable aspect of Viking commerce.

The majority of these slaves were exported to the Islamic world, particularly the Arab Caliphate. Slaves served various purposes beyond labour, including being used to settle debts and as sacrifices for religious ceremonies, which were commonplace during that era. The Vikings didn't hesitate to enslave their fellow compatriots. Additionally, they frequently raided regions such as Frisia, Ireland, and Slavic territories to capture slaves. However, a significant portion of the slave population also originated from North Africa due to Viking raids in that region.

Other Trading Goods

In addition to their diverse range of goods, the Vikings engaged in trade involving various other items. They exported wine from Germany, which symbolised wealth, and exported timber since pine wood was abundant in Scandinavia that remained after using it for longships. Before sugar became a prominent commodity, honey was also imported. During their travels, the Vikings participated in the trade of numerous goods like tin, wheat, wool, wood, iron, leather, fish, and walrus ivory. Furthermore, animal skins served as raw materials for local Viking artisans who created various items, including combs, needles, games, and decorations, from materials like whalebones, seal and walrus skins, and walrus tusks.

Lesson for Modern Businesses: Reducing Risk Through Diversification

Imagine the situation when the Vikings set sail for trade. They began their voyages with goods like amber, pine wood, luxurious fur, and even slaves to sell. Upon their return from journeys to distant lands, their ships were laden with valuable items such as Arabic silver, coins, fabrics, spices, silk, fruit, wine, and various other goods from the south. This is why they established a vast trading network that spanned most of the known world by the end of the Viking Age, creating a formidable trading empire. In conclusion, modern entrepreneurs can draw important lessons from their success:

The Vikings were savvy traders. They didn't put all their eggs in one basket, which would have been risky. Instead, they dealt with a range of goods, ensuring a steady income. The lesson here is to diversify your products in trade to reduce the risk of problems if the demand or supply for one item changes. For instance, if the supply of one product decreases, you can pivot to other goods without worrying about losing profit. In essence, the Vikings understood how diversification could mitigate risks related to fluctuations in demand or supply of specific products.

Indeed, the Vikings knew which products would yield the most profit with minimal investment. For instance, they capitalised on the abundance of pine trees and amber in their homeland. Through their efforts, they exported these items, reaping substantial profits.

In addition, the Vikings were adaptable traders who recognised market demands. Silver, for instance, became crucial as it was used for coinage. Realising the advantages of a currency-based system over barter, they actively traded for silver. This shift allowed them to buy and sell various products using silver coins, streamlining their trade and ensuring a stable income.

As concluded, Scandinavia's flourishing economy during the Viking Age, driven by silver, facilitated local trade and craftsmanship. They crafted a wide range of items from materials like whalebones, seal and walrus skins, and walrus tusks, all of which they traded alongside commodities such as honey, tin, wheat, wool, wood, iron, fur, leather, fish, and walrus ivory. While slave trading and weapon loot marred their activities, they also expanded their global trade network, acquiring goods like glass, amber, silk, spices, precious metals, jewellery, and textiles. Their diversified trading approach minimised risks and adapted to evolving markets, even pioneering the use of precious metals like silver for trade, leading to the introducing of a coinage system. In essence, modern businesses can draw valuable lessons from the Vikings' diversification and adaptability to maintain consistent profitability.

Chapter 7

ADAPTABILITY AND FLEXIBILITY

The transition from raiding to trading in the Viking society met a very successful response and kept the Vikings on top. Their economy was strong, much better than their immediate neighbours in Europe. A modern marketing team can learn much from the Viking culture regarding adaptability.

Viking Adaptability to Changing Market Demands and Trade Routes

The Vikings were ruthless traders and merchants, and they turned their skills to raiding new territories. A study of the Viking trade routes would reveal that they traded with other communities more than they feared for raids. Apart from helping the innocent, some Vikings were falsely accused of plundering, and their role in exploring and discovering world-changing aspects was well-preserved. As the Viking reign expanded, they encountered different cultures, and the markets changed. While they established themselves as successful traders of certain goods, their ability to adapt quickly to those changing markets and demands enabled them to continue to grow.

The shift from raiding to trading as a response to changing circumstances

Vikings were known for their violent raids in longboats designed for speed and navigation along rivers and shallow coastal waters.

But when the aggressive tactics of the Vikings became less profitable, they began to trade instead of raid. This shift was prompted by changing political conditions in Scandinavia and new opportunities in trading routes opened up by advances in shipbuilding technology. At the time, there was a lot of demand for the fur the Vikings had to offer, so they started selling it to buyers from other countries. They also traded with people from England and Russia.

The Viking Age seemed all about pillaging and looting. However, there was a lot more to their trading strategy than that. The Vikings were very adaptive and often changed their trading strategies to match the needs of their situation.

One example of this is how they adapted to the changing political climate in Europe. When Christianity became more prominent in Scandinavia, many Vikings converted to Christianity and stopped raiding so much. Instead, they started trading instead of raiding, which was one way they could continue making money without being seen as criminals in their home countries.

Another example is when they came across new cultures and saw the demand for certain products—such as slaves or ivory—the Vikings would take advantage of these opportunities by offering these items for sale if they did not have them already available in their inventory.

They did this because they realised they could make more money by expanding into new markets than by continuing to focus solely on Western Europe. It demonstrates that even though the Vikings were fierce warriors, they were also very adaptable people who knew how to make changes to succeed in whatever endeavour they pursued.

Staying Updated with Modern Market Trends and Regulations

You want to stay on top of modern market trends and regulations as a business. It can help you make informed decisions that will benefit your business, as well as help you avoid any costly mistakes.

Keeping up with the latest news relevant to your industry is essential, especially concerning regulation changes or new technological developments. You'll want to know about these things so that you can take advantage of them when they happen or plan for them and prepare your company for whatever comes next.

Exploration of Byzantine and Middle Eastern markets

In the early 900s CE, Scandinavian Viking traders began exploring the East. They established a permanent settlement in Constantinople—the capital of the Byzantine Empire—and traded with local merchants for silk and other luxury goods. The Vikings then sailed home with their loot, selling their goods at a premium price due to their scarcity.

The Vikings' success in Constantinople demonstrates how they adjusted their trading strategies based on the needs of their customers and the availability of resources in new markets, as well as how these changes affected prices for Viking goods.

The importance of understanding and complying with modern regulations

It is important to understand and comply with modern regulations. It would be best if you stayed up-to-date with current industry standards and best practices to provide the best possible service for your clients. It will help you avoid fines, penalties, and even criminal prosecution.

Strategies for Adapting to Changing Business Environments

Vikings adapted to changing business environments by developing new transportation, communication, and production methods. The first Viking ships were built between 500 and 300 BC, and they could travel across the sea due to their strong wooden hulls that could withstand the rough seas of the North Atlantic Ocean. As a result, they expanded their trade routes, leading them into contact with other peoples around Europe, such as the Franks, Slavs, and Frisians (Northumbria). By trading with these peoples, they could obtain new resources such as silver coins, which they used as currency for their society and later used in trade with countries like England and France.

Expeditions to Greenland and North AmeriCa

The Vikings, who lived in Scandinavia, survived the harsh winter months by hunting and fishing. In addition to these survival skills, they had developed a strong work ethic. They valued hard work and believed that those who did not work would not be longer. This attitude contributed to their ability to adapt to new trade routes and opportunities.

For example, Erik the Red was a Viking who discovered Greenland. He was born around 970 and lived in Iceland during the 10th century. He was exiled from Iceland by his father, and he went to live in Norway, where he met his wife Thjodhild, a woman who had been kidnapped by Vikings and taken away as a slave.

Following this, Erik wanted to explore new territories, so he travelled with his family to Iceland, where they stayed for three years before going on their first expedition to Greenland. They explored these lands for two years before returning to Iceland, where they had two children: Leif Erikson, who would later become an explorer himself, and Thorvald Erikson.

Erik's son, Leif Erikson, was also an explorer. He sailed west in 986 CE and discovered North America, which he named Vinland (meaning "land of grapes"). The Vikings did not settle there permanently because they needed help to grow crops there.

How Technology and Innovation Played a Role in Viking Adaptability

The Vikings travelled worldwide, bringing their culture and way of life to new lands. They were adaptable and could use technology and innovation as they encountered different environments and obstacles.

One example of this was the evolution of their ship designs. When they first began sailing, they used longships with a single sail on each side. As they moved into colder climates, they developed double-sailed ships with a large central sail that could be lowered in the windy conditions found at sea. In addition to this innovation, the Vikings created larger boats that could carry more passengers and cargo over longer distances.

Another example of Viking adaptability comes from their use of technology related to navigation by stars or sun position during long voyages across open water where few landmarks were available for reference points. It allowed them to navigate safely even when visibility was poor due to fog or other weather conditions such as storms/hurricanes/typhoons, etc.

Lesson for Modern Businesses

Vikings were known for their adaptability, which allowed them to survive in a vast range of climates, from the frozen North to the stifling heat of the Mediterranean. They were also flexible in their approach to warfare, allowing them to adapt their tactics depending on their opponent and environment.

The importance of adaptability and flexibility in today's global market

The modern world is full of opportunities for entrepreneurs to thrive, but it also presents obstacles that can make it difficult for businesses to grow. For example, what if your business is based in a country with stringent labour laws? Or what if you're selling products or services in multiple countries, but different laws govern sales in each country?

If you're an entrepreneur trying to build a business in today's marketplace, you must understand how the Vikings' adaptability and flexibility helped them flourish in their time. You might be surprised at how valuable these traits can be when applied to modern businesses!

Businesses should stay agile and responsive to changing conditions

If a company cannot adapt when needed, it will fall behind its competitors who do have this ability. Therefore, companies must be willing to try new things even if they don't work out immediately or ever! It means being open-minded about new ideas and implementing them quickly if they prove successful. Given current circumstances, it also means letting go of traditional practices that aren't working or effective. It's one of the things that make Vikings so great.

Real-Life Examples of Adaptability

Indeed, as relevant today as a millennium ago, the Viking trading techniques and principles are invaluable to anyone dealing with change in a volatile marketplace.

Modern companies that successfully adapted to market changes

One example is Netflix, which started as a DVD-by-mail service in 1997 but grew into an online streaming service worth over $100 billion today. Another example is Amazon, which started as an online bookstore but now offers everything from diapers to cloud computing services.

These two companies show us that adaptability and flexibility can be powerful tools in business—but they're not just for startups. Large, established corporations like Walmart constantly seek ways to innovate and keep up with their customers' changing needs.

What Advantages Do Amazon Get for Its Adaptability

Amazon also started small but quickly grew into an online retail powerhouse due to its adaptability. The company responded quickly when Apple released the iPad by developing an app that allowed users to browse books on their phones instead of having to go directly through Amazon's website or store locations; this helped keep people engaged with their brand even when shopping away from home (or at least when they had their phone with them).

Viking society was built upon the idea of a strong community. The Vikings' ability to adapt— to new cultures, languages, and technology—made them successful as traders, explorers, and conquerors.

Today, this kind of flexibility is essential in business. It doesn't matter if you're in sales or marketing: your company will succeed if you adapt quickly to changing trends and circumstances.

In essence, if we believe the research and observe the Vikings historical evidence we have examined, then there are real parallels we can draw for modern trading and business leaders on how to deal with continually changing conditions. They must remain keenly aware of what is happening and adapt to new environments and market conditions.

In an age of ever-changing technology, the most successful and resilient businesses will embrace change and adapt to new regulations. By being flexible and understanding consumers' ever-evolving needs and desires, business leaders can help their companies prosper in a competitive marketplace.

Chapter 8

GLOBAL NETWORKING AND COLLABORATION

Vikings were skilled seafarers and traders known for their proactive approach to international connections and collaboration in the Viking Age. They were involved in trade exploration, which helped them connect with various cultures and people. They also established many trade routes from the Baltic to the Mediterranean Sea. Vikings traded goods like precious metals, fur, amber, etc. The exchange of these commodities enhanced trade and fostered cultural exchange between diverse groups.

Additionally, the Vikings' exploration and colonisation efforts helped them establish settlements. The settlements were in regions like Iceland, Greenland, Island, England, and Vinland. So, these settlements involved collaboration and direction with the local population. The proactive nature of Viking International's collaboration and connection was evident from their efforts. They were known as people who raided the coastal regions and significantly impacted different areas they encountered.

Erik's Collaboration with Foreign Traders

Erik, whose story we discussed in Chapter 7, was a Viking merchant who exemplified the strategic approach to international collaboration and connection. In the Birka, a Viking merchant, Erik, was known for his business skills and willingness to collaborate with foreign traders. His business story exemplifies the Viking's adventurous spirit about expanding the trading networks in the summer. Erik got an idea and decided to look for foreign traders and establish new collaborations. He knew that diversification was the main key to success in commerce. So, he came from Southward and navigated the waterways of the Baltic Sea.

During his journey, he encountered many merchants from the Byzantine Empire, Abbasid Caliphate, and other foreign lands. He was involved in trade and barter systems and exchanged many Scandinavian goods, including treasures, spices, silk, luxury items, etc. His open-minded approach spread quickly, and traders from different lands arrived in Birka. Due to his collaboration with the foreign traders, he not only amassed much wealth but also enriched the culture. The town becomes a central point for merging different languages, customs, and traditions. It was his proactive approach to the international connection.

Another anecdote that highlights the thriving Viking trade to Byzantium is the story of the success of international partnerships. There was an exotic marketplace in Constantinople that was the main place for the Byzantine Empire. This marketplace symbolised trade connections between the Vikings and the Byzantines. The main key figure in this trade network was a Viking merchant. He was known for adventurous spirits beyond the Scandinavian regions in search of foreign people. His story depicts International partnerships and prosperities. He had navigated the waters and established a connection with the Byzantines. Through negotiation and diplomatic relations, he secured trading opportunities in Constantinople.

Benefits of Global Networking and Collaboration

Global networking among Vikings has brought several advantages to Scandinavian people. These trade networks stretched from the Baltic region to the Mediterranean region, which helped them support economic growth. Moreover, they also interacted with people of diverse cultures, which helped them reach cultural exchange through their global connections. Vikings have gained access to new techniques and technologies, including agriculture, metalwork and shipbuilding. They have also established many diplomatic relationships with the foreign people. The working settlement also allowed them for strategic expansion and control of trade routes. As Vikings explored new territories, it helped them acquire knowledge about different fields, which facilitated their exploration for the groundwork of future discoveries.

In the heart of the Viking Age was a trader named Bjorn. He had a great thirst for adventure and exploration of different cultures. He had entered the Abbasid caliphate market, where he encountered traders from diverse regions along the Silk Road. This Vikings trader got many advanced navigation tools. Additionally, he has also navigated through seas using an astrolabe, which guided him in the night. He also reached distant Islands and established a trading post there. Along with this, he has also gained access to different resources like textiles and spices. His partnership with another trader, Nasir, exemplified the dissemination of knowledge on distant shores.

Strategies for Fostering Modern Global Networking and Collaboration

By taking inspiration from Viking strategies, modern businesses can easily adopt key global networking and collaboration principles. As Vikings were fearless explorers, modern organisations should also explore new markets, trends, and opportunities. They must encourage a culture of willingness for economic growth and success of the territory. Vikings also adapted to the customs and practices of the areas they encountered. Similarly, modern businesses should also be adaptive and flexible to the unique legal, cultural, and economic landscape of different regions. Moreover, Vikings established partnerships and alliances. It was for mutual benefit. So modern businesses should also adopt these strategic partnerships to leverage the strength to achieve mutual goals.

Vikings always carefully assess all the risks before going on the journey. Similarly, modern organisations also conduct thorough risk assessments when entering new markets. Whenever they address legal, financial, or other operations, they must adhere to all the risks before starting their partnership with any other company.

One of the most prominent examples is a multinational corporation known as NordiTech. This company has also drawn inspiration from Viking practices and strategies in expanding its global presence. The company embarked on establishing partnerships with traders from diverse regions. They started with thorough research of the potential partners and the key market. The team has been so impressed with digital connectivity.

They have also invested in cultural sensitivity training to ensure their approach is beneficial. NordiTech representatives have also travelled to distant countries to build connections with foreign traders. Through fearless exploration and commitment to long-term growth, NordiTech has taken a very fruitful approach. The global presence of this company increases significantly and fosters economic growth.

Lesson for Modern Businesses

In this modern and fast-paced business world, fostering collaboration and global networking is crucial. The lessons learned from Vikings can guide contemporary modern businesses to success.

Adopting Cross-Cultural Thinking:

As the Vikings ventured far from their homeland, modern businesses should explore new markets and trends beyond their borders. This marketplace offers potential for innovation, technology, and growth. Understanding cultural differences and respecting those differences is paramount. So modern businesses should also invest in cross-cultural thinking and training for navigating international partnerships. It promotes effective communication and trust between the two parties.

Making Strategic Partnerships

Forming strategic partnerships with other organisations worldwide can provide easy access to customers, new markets, and resources. Modern businesses should also attend international trade conferences to connect with potential partners and stay updated with trade and industry development. All these events will prove them valuable in global networking opportunities. Similarly, the Vikings have also adopted this strategy for success in global networking.

Utilising Digital Tools

By embracing digital tools and platforms to facilitate global collaboration, modern businesses can have a competitive edge in the market. Project management software, video conferencing, and data analytics are some communication and decision-making tools that will benefit them. So, as Vikings were adaptable in all their operations, the modern businesses should also adapt to the strategies based on market conditions and feedback from the foreign partners.

Successful Vikings Global Networks

The Viking Age occurred a thousand years ago. There are numerous real-life examples of successful Viking global networking and connections with other cities. These examples show how Vikings traded and explored different places.

Varangian Guard:

The Varangian Guard is a notable example of Viking Global networking. It was an allied group of warriors of which the Vikings were also the members. They acted as personal bodyguards of the Byzantine Emperors. This was helpful for them in making close collaboration with the Byzantine Empire. It inspired the relationship of Vikings with the diplomats at that time. So, this is a great example of Vikings contributing to the global collaboration.

Vinland Settlement:

Vinland was a part of North America where the Viking trader Leif Erikson built a settlement. It was around the 11th century when he demonstrated the ability of Vikings to establish colonies in distant regions. This Vinland settlement showed their strategy of establishing and exploring distant lands.

Trade Routes to Byzantine Empire:

Vikings also established trade routes that reached the Byzantine Empire and the Abbasid caliphate. These trade routes exchanged goods, including silver, amber, and copper metals. It was also used to exchange valuable commodities, including precious metals, luxury items, spices, and silk. These trade routes are the main examples of successful Viking global interactions.

In conclusion, the Viking's success during the Viking Age in global collaboration and networking has many lessons for modern businesses. Modern businesses should actively establish international relations, attend conferences, and collaborate for a successful global marketplace. They should be open to adapting strategies and embracing digital tools and platforms. By leveraging these lessons, modern businesses can easily embrace the Viking spirit of collaboration, exploration, and adaptation for success in the globalised world.

Chapter **9**

RISK MANAGEMENT

Throughout history, many examples of cultures' awareness of the risks related to trading exist. As we discussed the Viking shipbuilding techniques in Chapter 1, let us enlighten the advantages of trade and their risk management strategies now! The Vikings traded with many cultures, and as traders themselves, they faced different types of business risks. The main goal of their organisation was to minimise these risks that could affect their lives, their voyages, and even their overseas territory. At the same time, they would try to increase their profits by not falling into unexpected accidents that could appear during a long journey.

Viking risk management practices, including insurance and security measures

The Norsemen knew how important it was to protect themselves from harm during battle or while travelling on land or sea. They used many strategies to prevent injury or death in case something went wrong during a battle or journey across long distances without modern-day technology such as cars or planes available today!

Let us take an example:

In 873 CE, Danish king Sigfred's emissaries travelled to Worms to agree with King Louis the German. They aimed to secure peace in the border areas between the Danes and the Saxons' territories. Merchants would, therefore, be able to trade in peace and safety in the neighbouring empire.

The emissaries had come equipped with a charter or letter of safe conduct issued by their king, which guaranteed they could travel through territories under his control without being harassed or prosecuted. The charter also served as proof of authorisation to engage in business activities.

Lessons from Viking Risk Management

The Vikings were the first to make a name for themselves as traders and seafarers. While they were known for their prowess in battle, they were also known for their ability to navigate the seas and trade with other cultures.

But how did they manage risk? How did they stay safe and productive when venturing out into the unknown? The answer is simple: proactive risk management.

Importance of proactive risk management in shipping and trading operations

A proactive approach to risk management involves identifying potential risks before they occur. It can be done by identifying potential hazards, evaluating those hazards, and then determining ways to prevent them from happening.

When it comes time to trade with another culture or sail across the ocean, you can't wait until disaster strikes before acting—you have to plan and prepare yourself for any scenario. It means thinking about what could go wrong (and how you'll deal with it) before it happens!

Businesses should develop comprehensive contingency plans and security measures

The Viking Age was a time of unprecedented opportunity and risk. At that time, kings were expected to protect their people. It was especially true in the case of trade when agreements between countries were a necessary part of doing business.

Trading agreements are an example of this protection. They testify to peaceful coexistence between countries and the fact that trade could not occur without such agreements and protection from kings.

Today, there is a lot of talk about security—but what does it mean to be secure? Security can mean different things to different people when you think about it. For some people, security is a feeling of being physically safe; for others, it's about financial stability; and for others, it is about reputation or reputation management.

For any company to achieve security and protection from various forms of harm or damage—whether physical or financial—they need an effective plan that accounts for all possible risks. It means having procedures that protect against internal and external threats and monitoring systems that quickly identify vulnerabilities before they become problems.

In addition to these measures, businesses should also ensure they're practising good cyber hygiene by keeping up-to-date software on all devices used by employees and implementing strong passwords.

Historical Viking risk management VS modern business practices

The way that the Vikings managed risk is often cited as one of the reasons they were able to conquer so much of Europe. They had an uncanny ability to plan and think through all the potential outcomes of each action they took—and then take steps to mitigate failure.

And it's not just their success that makes these strategies so fascinating: it's how closely they resemble modern business practices. The following lessons can help you apply Viking-style thinking to your company's day-to-day operations.

The Vikings had a variety of ways in which they managed the risks involved with exploration. For example, they would often set up certain rules that everyone in the group had to follow, such as having one person stay behind at all times to watch over the rest of them. They also took time before setting off on an expedition to ensure everything was in order beforehand—this included ensuring that everyone had enough supplies and getting any necessary repairs done before leaving port.

Implementing Modern Risk Management Practices

In the modern day and age, it has become necessary to implement a robust risk management strategy. This is because there are many challenges facing businesses today. These challenges range from economic downturns, competition, and natural disasters. To overcome these challenges, companies should implement a risk management strategy that will help them mitigate their risks.

Thus, the risks involved with modern businesses are varied and complex, requiring a sophisticated approach to managing them.

Risk management strategies that are effective in today's business climate include:

- Identifying risks early and developing plans for their prevention or mitigation.

- Implementing employee training programs that cover how to identify and report potential risks.

- Developing strategies for keeping up with changing regulations and laws, which can significantly impact a company's ability to operate safely and effectively.

- Working closely with insurance companies to ensure the right coverage is in place.

Strategies for modern businesses to implement effective risk management practices

Risk management has become more important than ever, especially as organisations face increasing pressure to innovate and compete. With the right strategies, modern businesses can implement effective risk management practices that help them stay ahead of the curve and proactively deal with potential dangers.

Recognise risks early

Risk management isn't just about avoiding potential pitfalls—it's also about identifying opportunities to help you grow your business. For example, if you know that a big new market is opening up for your business, you'll want to be ready to seize the opportunity before anyone else does!

Conduct a Risk Audit

A risk audit involves identifying all the potential risks that your company faces and then determining how those risks will impact your business. It would help if you also considered what can be done to mitigate each risk and how much it would cost you to do so.

Know what you're up against

Knowing what risks will be most dangerous for your business can make you better prepared. You can also use this information when deciding where best to invest your time and resources so that they're used most effectively in managing these risks down the road.

Identify Your Most Vulnerable Assets

Your most vulnerable assets are those that are most important to your company's success and growth, such as data and intellectual property; human capital (employees); physical capital (buildings); financial capital (cash flow); operational capital (production equipment); and reputation capital (brand equity). By identifying these assets early in the process, you can better protect them from any potential threats or losses resulting from cyberattacks or other incidents that could negatively impact their security/availability/integrity/confidentiality/privacy (SCIP).

Remember people-related risks!

People are probably the most important asset in creating value in any organisation—so don't overlook them when thinking about what could go wrong with your organisation's current processes or systems!

Create a Comprehensive Emergency Response Plan

Having an emergency response plan can make all the difference in a crisis. The plan should include how you will respond to an emergency and communicate with your employees and customers during the crisis. This plan must be written down and distributed to all employees so they know their responsibilities in an emergency.

Real-Life Examples of Effective Risk Management

Risk management practices utilised by Amazon.com have been so successful that they have made the company one of the most valuable in the world. They've done this by focusing on minimising risk while maximising profit potential. For example, they use technology to minimise human error and maximise efficiency, reduce inventory costs by allowing customers to order directly from manufacturers, and minimise sales costs through discounts on shipping charges when customers buy more than one item.

So, there are many lessons to learn from the Vikings and their experience managing risk. As long as people make trades on any scale, they could encounter setbacks. When businesses practice good risk management, their overall chances of success increase exponentially; without it, disaster is certain. Protecting your crew or cargo from possible threats is especially crucial in a society increasingly relying on international trade and imports. By increasing their savings and reducing their debt load, a habit that would likely have mirrored the financial practices of the Norse traders, individuals can also reduce their risk of economic catastrophe arising from events like extreme illness or property loss due to fire or theft. Even the Vikings knew this—and now you can too!

CONCLUSION

Vikings ruled the sea for almost four centuries, and their brave souls led them to the faraway sea of the Middle East, which was quite a feat at that time. They did not only polish their ships but polished their crafts into fine shipbuilding and themselves as tremendous traders. There are several ships in different Norway museums to see their fine craftsmanship regarding shipbuilding. But what we can't see is the lesson they left for us.

In essence, the Vikings' rich maritime and trade legacy offers a treasure trove of lessons encompassing adept seamanship, astute market assessment, tactical location selection, ethical trade conduct, flexibility, and astute risk management. By delving into the ocean of Viking history, contemporary shipping and trading enterprises can enhance their endeavours, boost efficiency, and bolster their bottom line in the complex landscape of today's global markets. The Vikings' historical prowess in these areas serves as an enduring source of wisdom for modern businesses seeking to thrive in an ever-evolving world of commerce.

FINAL SUMMARY

When you hear about Vikings, what are some of the things that come to your mind first? Regardless of the fascinating ideas you may have come up with, they are, in fact, more charismatic than you think.

The 8th to 11th century saw more Vikings taking to the sea due to the lack of economic opportunities and a strong desire for adventure. While they were classed as shrewd for the most part, this particular characteristic helped them to establish healthy foreign relationships. In turn, they received protection from kings and other royals.

When it came to shipbuilding and design, the Vikings were in a class of their own. The intricate details and use of fresh timber made it easier for them to design sea-worthy vessels, such as the Gokstad and Roskilde 6, that could navigate shallow water with minimal difficulty. Their ability to create vital ports along existing trade routes ultimately increased commerce and cultural exchange.

The Vikings possessed an unwavering approach to honest dealings and fair pricing, evident in their proactive risk management skills in identifying and mitigating potential hazards. Delving into the ocean of Viking history reveals that their prowess in the previously mentioned areas serves as an enduring source of wisdom for modern businesses seeking to thrive in an ever-evolving world of commerce.